Project MIND
Math Is Not Difficult

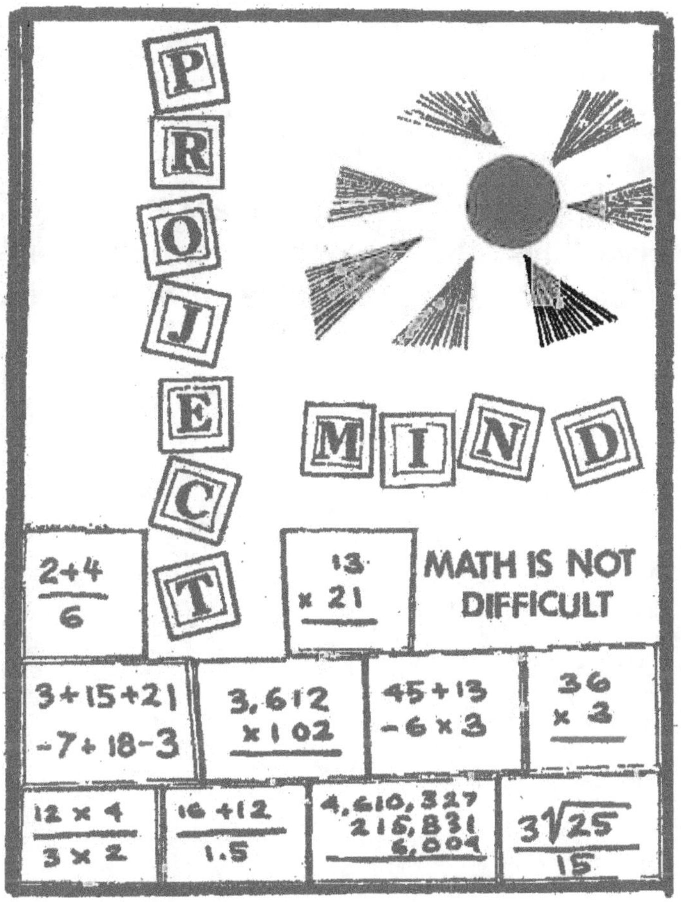

First Grade
Mental Math Flash Cards

Hui Fang Huang "Angie" Su, Ed.D.
Project MIND, Inc.

The Mental Math Game

The students form two teams and come up to the bells two at a time. Upon looking at the math problem on a yellow card, they solve the problem mentally as fast as they can, usually within three seconds. The winner continues on while the loser moves to the end of his line. To be an intermediate champion, one must respond to three problems in a row correctly. After four intermediate champions are picked (depending on the size of your group, you must make sure that each student had a t least three chances), they are then entered into the second level of competitions with the green cards (more difficult problems.) To be a runner up for the grand champion title, competitors must also respond correctly to three problems in a row. Two runner-ups for the advanced level cards (red) are picked. They now compete for the title. The first person to respond to three red card problems in a row correctly is the grand champion.

Variations:

- The students compete in four areas: Mentally solve math problems with cards (visual aids), mentally solve math problems without cards, word problems, and equations (a string of problems to solve as the reader reads them.)
- The game can be played with your own class, another class, your grade level, or with other grade levels (fourth grade competing against fifth grade, third grade competing against fourth grade, etc.)
- If you have an advanced group, make sure that they use the cards for the next grade.
- Decimals and fractions can be added for third through fifth grade.

Pre-Kindergarten/Kindergarten:

- Level 1 – Yellow Cards: Number identification and shape identification
- Level 2 – Green Cards: Number identification (up to 100), identify the missing number, and adding and subtracting up to 5.
- Level 3 – Red Cards: number sequencing, and adding and subtracting up to 10
- Equations: strings of numbers which add and subtract up to 10
- Word problems: Simple one step, how many items? Adding or subtracting up to 10

First Grade:

- Level 1 – Yellow Cards: Adding and subtracting numbers up to 10
- Level 2 – Green Cards: Adding and subtracting two-digit numbers and adding three digit numbers
- Level 3 – Red Cards: Adding and subtracting three-digit numbers

Second Grade:

- Level 1 – Yellow Cards: Adding and subtracting two-digit numbers

- Level 2 – Green Cards: Adding and subtracting two-digit numbers with carrying and borrowing, and multiplication and division facts
- Level 3 – Red Cards: Adding and subtracting three-digit numbers with carrying and borrowing; two-digit multiplication

Third Grade:

- Level 1 – Yellow Cards: Adding and subtracting two-digit numbers with carrying and borrowing; single digit multiplication and division
- Level 2 – Green Cards: Adding and subtracting three-digit numbers with carrying and borrowing, and two-digit multiplication and division
- Level 3 – Red Cards: Adding and subtracting four-digit numbers with carrying and borrowing; three-digit multiplication and division

Fourth Grade:

- Level 1 – Yellow Cards: Adding, subtracting, multiplying, and dividing fourth grade level problem
- Level 2 – Green Cards: Adding, subtracting, multiplying, and dividing fourth grade level problems that are harder than Level 1
- Level 3 – Red Cards: Adding, subtracting, multiplying, and dividing multi-digit fifth grade level problems

Fifth Grade:

- Level 1 – Yellow Cards: Adding, subtracting, multiplying, and dividing fifth grade level problem
- Level 2 – Green Cards: Adding, subtracting, multiplying, and dividing fifth grade level problems that are harder than Level 1
- Level 3 – Red Cards: Adding and subtracting six digit numbers with carrying and borrowing, and multiplying and dividing multi-digit problems

$$\begin{array}{r} 1 \\ +\ 0 \\ \hline \end{array}$$

$$\begin{array}{r} 0 \\ +\ 1 \\ \hline \end{array}$$

$$\begin{array}{r} 0 \\ +\ 1 \\ \hline 1 \\ \hline\hline \end{array}$$

Project MIND
First - Yellow

$$\begin{array}{r} 1 \\ +\ 0 \\ \hline 1 \\ \hline\hline \end{array}$$

Project MIND
First - Yellow

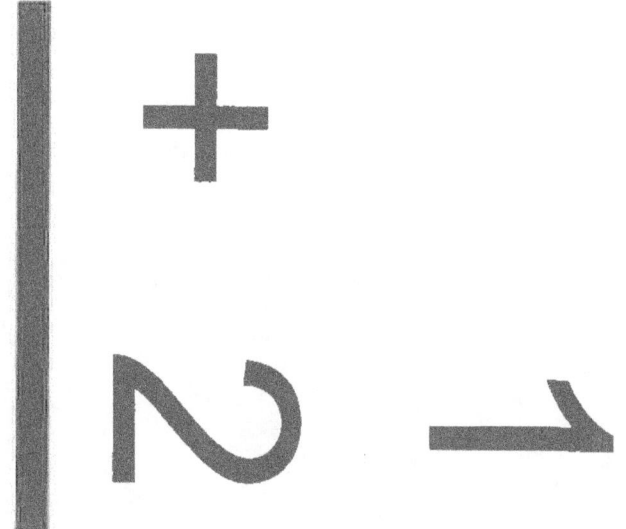

$$\begin{array}{r} 1 \\ +\ 2 \\ \hline \end{array}$$

$$\begin{array}{r} 1 \\ +\ 1 \\ \hline \end{array}$$

$$\begin{array}{r} 1 \\ +\ 1 \\ \hline 2 \end{array}$$

Project MIND
First - Yellow

$$\begin{array}{r} 1 \\ +\ 2 \\ \hline 3 \end{array}$$

Project MIND
First - Yellow

$$\begin{array}{r} 1 \\ +\ 5 \\ \hline \end{array}$$

$$\begin{array}{r} 1 \\ +\ 4 \\ \hline \end{array}$$

$$\begin{array}{r} 1 \\ +\ 4 \\ \hline 5 \\ \hline\hline \end{array}$$

Project MIND
First - Yellow

$$\begin{array}{r} 1 \\ +\ 5 \\ \hline 6 \\ \hline\hline \end{array}$$

Project MIND
First - Yellow

$$\begin{array}{r} 2 \\ +\ 0 \\ \hline \end{array}$$

$$\begin{array}{r} 1 \\ +\ 6 \\ \hline \end{array}$$

$$\begin{array}{r} 1 \\ + \quad 6 \\ \hline 7 \\ \hline \hline \end{array}$$

Project MIND
First - Yellow

$$\begin{array}{r} 2 \\ + \quad 0 \\ \hline 2 \\ \hline \hline \end{array}$$

Project MIND
First - Yellow

$$\begin{array}{r} 2 \\ +\ 3 \\ \hline \end{array}$$

$$\begin{array}{r} 2 \\ +\ 1 \\ \hline \end{array}$$

$$\begin{array}{r} 2 \\ +\ 1 \\ \hline 3 \end{array}$$

Project MIND
First - Yellow

$$\begin{array}{r} 2 \\ +\ 3 \\ \hline 5 \end{array}$$

Project MIND
First - Yellow

$$\begin{array}{r} 3 \\ +\ 0 \\ \hline \end{array}$$

$$\begin{array}{r} 2 \\ +\ 4 \\ \hline \end{array}$$

$$\begin{array}{r} 2 \\ +\ 4 \\ \hline 6 \\ \hline\hline \end{array}$$

Project MIND
First - Yellow

$$\begin{array}{r} 3 \\ +\ 0 \\ \hline 3 \\ \hline\hline \end{array}$$

Project MIND
First - Yellow

$$\begin{array}{r} 3 \\ +\ 2 \\ \hline \end{array}$$

$$\begin{array}{r} 3 \\ +\ 1 \\ \hline \end{array}$$

$$\begin{array}{r} 3 \\ +\ 1 \\ \hline 4 \\ \hline \end{array}$$

Project MIND
First - Yellow

$$\begin{array}{r} 3 \\ +\ 2 \\ \hline 5 \\ \hline\hline \end{array}$$

Project MIND
First - Yellow

$$\begin{array}{r} 3 \\ +\ 5 \\ \hline \end{array}$$

$$\begin{array}{r} 3 \\ +\ 3 \\ \hline \end{array}$$

$$\begin{array}{r} 3 \\ +\ 3 \\ \hline 6 \\ \hline \end{array}$$

Project MIND
First - Yellow

$$\begin{array}{r} 3 \\ +\ 5 \\ \hline 8 \\ \hline \end{array}$$

Project MIND
First - Yellow

$$\begin{array}{r} 4 \\ +\ 2 \\ \hline \end{array}$$

$$\begin{array}{r} 4 \\ +\ 1 \\ \hline \end{array}$$

$$\begin{array}{r} 4 \\ +\ 1 \\ \hline 5 \end{array}$$

Project MIND
First - Yellow

$$\begin{array}{r} 4 \\ +\ 2 \\ \hline 6 \end{array}$$

Project MIND
First - Yellow

$$\begin{array}{r} 5 \\ +\;1 \\ \hline \end{array}$$

$$\begin{array}{r} 4 \\ +\;3 \\ \hline \end{array}$$

$$\begin{array}{r} 4 \\ +\ 3 \\ \hline 7 \\ \hline \end{array}$$

Project MIND
First - Yellow

$$\begin{array}{r} 5 \\ +\ 1 \\ \hline 6 \\ \hline\hline \end{array}$$

Project MIND
First - Yellow

$$\begin{array}{r} 6 \\ +\ 2 \\ \hline \end{array}$$

$$\begin{array}{r} 5 \\ +\ 2 \\ \hline \end{array}$$

$$\begin{array}{r} 5 \\ +\ 2 \\ \hline 7 \end{array}$$

Project MIND
First - Yellow

$$\begin{array}{r} 6 \\ +\ 2 \\ \hline 8 \end{array}$$

Project MIND
First - Yellow

$$\begin{array}{r} 2 \\ -\ 1 \\ \hline \end{array}$$

$$\begin{array}{r} 2 \\ -\ 0 \\ \hline \end{array}$$

$$\begin{array}{r} 2 \\ -\ 0 \\ \hline 2 \end{array}$$

Project MIND
First - Yellow

$$\begin{array}{r} 2 \\ -\ 1 \\ \hline 1 \end{array}$$

Project MIND
First - Yellow

$$\begin{array}{r} 3 \\ -\ 0 \\ \hline \end{array}$$

$$\begin{array}{r} 2 \\ -\ 2 \\ \hline \end{array}$$

$$\begin{array}{r} 2 \\ -\ 2 \\ \hline 0 \end{array}$$

Project MIND
First - Yellow

$$\begin{array}{r} 3 \\ -\ 0 \\ \hline 3 \end{array}$$

Project MIND
First - Yellow

$$\begin{array}{r} 3 \\ -\ 2 \\ \hline \end{array}$$

$$\begin{array}{r} 3 \\ -\ 1 \\ \hline \end{array}$$

$$\begin{array}{r} 3 \\ -\ 1 \\ \hline 2 \end{array}$$

Project MIND
First - Yellow

$$\begin{array}{r} 3 \\ -\ 2 \\ \hline 1 \end{array}$$

Project MIND
First - Yellow

4 - 1 ———	3 - 3 ———

$$\begin{array}{r} 3 \\ -\ 3 \\ \hline 0 \end{array}$$

Project MIND
First - Yellow

$$\begin{array}{r} 4 \\ -\ 1 \\ \hline 3 \end{array}$$

Project MIND
First - Yellow

$$\begin{array}{r} 5 \\ -\ 0 \\ \hline \end{array}$$

$$\begin{array}{r} 4 \\ -\ 2 \\ \hline \end{array}$$

$$\begin{array}{r} 4 \\ -\ 2 \\ \hline 2 \end{array}$$

Project MIND
First - Yellow

$$\begin{array}{r} 5 \\ -\ 0 \\ \hline 5 \end{array}$$

Project MIND
First - Yellow

$$\begin{array}{r} 5 \\ -\ 2 \\ \hline \end{array}$$

$$\begin{array}{r} 5 \\ -\ 1 \\ \hline \end{array}$$

$$\begin{array}{r} 5 \\ -\ 1 \\ \hline 4 \end{array}$$

Project MIND
First - Yellow

$$\begin{array}{r} 5 \\ -\ 2 \\ \hline 3 \end{array}$$

Project MIND
First - Yellow

$$\begin{array}{r} 6 \\ -\ 1 \\ \hline \end{array}$$

$$\begin{array}{r} 5 \\ -\ 4 \\ \hline \end{array}$$

$$\begin{array}{r} 5 \\ -\ 4 \\ \hline 1 \end{array}$$

Project MIND
First - Yellow

$$\begin{array}{r} 6 \\ -\ 1 \\ \hline 5 \end{array}$$

Project MIND
First - Yellow

$$\begin{array}{r} 6 \\ -\ 3 \\ \hline \end{array}$$

$$\begin{array}{r} 6 \\ -\ 2 \\ \hline \end{array}$$

$$\begin{array}{r} 6 \\ -\ 2 \\ \hline 4 \end{array}$$

Project MIND
First - Yellow

$$\begin{array}{r} 6 \\ -\ 3 \\ \hline 3 \end{array}$$

Project MIND
First - Yellow

$$\begin{array}{r} 6 \\ -\ 6 \\ \hline \end{array}$$

$$\begin{array}{r} 6 \\ -\ 5 \\ \hline \end{array}$$

$$\begin{array}{r} 6 \\ -\ 5 \\ \hline 1 \end{array}$$

Project MIND
First - Yellow

$$\begin{array}{r} 6 \\ -\ 6 \\ \hline 0 \end{array}$$

Project MIND
First - Yellow

$$\begin{array}{r} 7 \\ -\ 6 \\ \hline \end{array}$$

$$\begin{array}{r} 7 \\ -\ 1 \\ \hline \end{array}$$

$$\begin{array}{r} 7 \\ -\ 1 \\ \hline 6 \end{array}$$

Project MIND
First - Yellow

$$\begin{array}{r} 7 \\ -\ 6 \\ \hline 1 \end{array}$$

Project MIND
First - Yellow

$$\begin{array}{r} 8 \\ -\ 1 \\ \hline \end{array}$$

$$\begin{array}{r} 7 \\ -\ 7 \\ \hline \end{array}$$

$$\begin{array}{r} 7 \\ -\ 7 \\ \hline 0 \end{array}$$

Project MIND
First - Yellow

$$\begin{array}{r} 8 \\ -\ 1 \\ \hline 7 \end{array}$$

Project MIND
First - Yellow

$$\begin{array}{r} 2 \\ +\ 9 \\ \hline \end{array}$$

$$\begin{array}{r} 2 \\ +\ 8 \\ \hline \end{array}$$

$$\begin{array}{r} 2 \\ +\ 8 \\ \hline \boxed{10} \end{array}$$

Project MIND
First - Green

$$\begin{array}{r} 2 \\ +\ 9 \\ \hline 11 \end{array}$$

Project MIND
First - Green

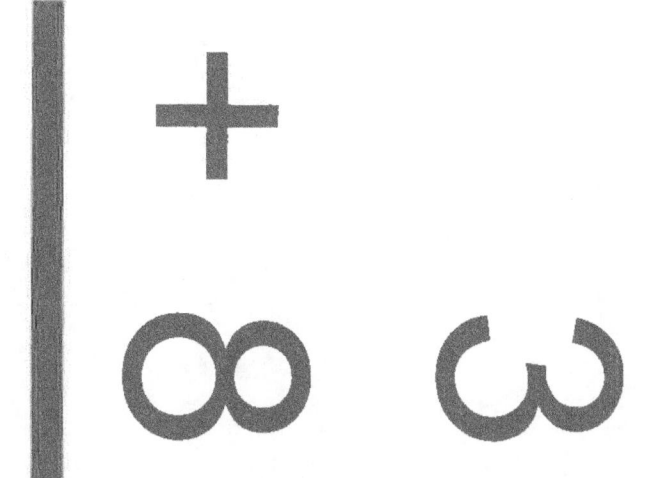

$$\begin{array}{r} 3 \\ +\ 8 \\ \hline \end{array}$$

$$\begin{array}{r} 3 \\ +\ 6 \\ \hline \end{array}$$

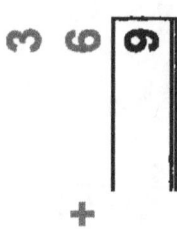

$$\begin{array}{r} 3 \\ + \; 6 \\ \hline \boxed{9} \end{array}$$

Project MIND
First - Green

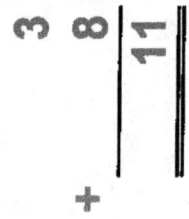

$$\begin{array}{r} 3 \\ + \; 8 \\ \hline 11 \end{array}$$

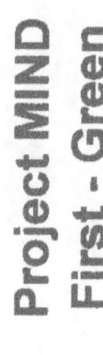

Project MIND
First - Green

$$\begin{array}{r} 4 \\ +\ 8 \\ \hline \end{array}$$

$$\begin{array}{r} 3 \\ +\ 9 \\ \hline \end{array}$$

$$\begin{array}{r} 3 \\ +\ 9 \\ \hline 12 \end{array}$$

Project MIND
First - Green

$$\begin{array}{r} 4 \\ +\ 8 \\ \hline 12 \end{array}$$

Project MIND
First - Green

$$\begin{array}{r} 5 \\ +\ 6 \\ \hline \end{array}$$

$$\begin{array}{r} 5 \\ +\ 5 \\ \hline \end{array}$$

$$\begin{array}{r} 5 \\ +\ 5 \\ \hline \boxed{10} \end{array}$$

Project MIND
First - Green

$$\begin{array}{r} 5 \\ +\ 6 \\ \hline 11 \end{array}$$

Project MIND
First - Green

$$\begin{array}{r} 6 \\ +\ 3 \\ \hline \end{array}$$

$$\begin{array}{r} 5 \\ +\ 7 \\ \hline \end{array}$$

$$\begin{array}{r} 5 \\ +\ 7 \\ \hline \boxed{12} \end{array}$$

Project MIND
First - Green

$$\begin{array}{r} 6 \\ +\ 3 \\ \hline 9 \end{array}$$

Project MIND
First - Green

$$\begin{array}{r} 6 \\ +\ 6 \\ \hline \end{array}$$

$$\begin{array}{r} 6 \\ +\ 4 \\ \hline \end{array}$$

$$\begin{array}{r} 6 \\ +\ 4 \\ \hline \boxed{10} \end{array}$$

Project MIND
First - Green

$$\begin{array}{r} 6 \\ +\ 6 \\ \hline 12 \end{array}$$

Project MIND
First - Green

$$\begin{array}{r} 7 \\ +\ 4 \\ \hline \end{array}$$

$$\begin{array}{r} 6 \\ +\ 7 \\ \hline \end{array}$$

$$\begin{array}{r} 6 \\ +\ 7 \\ \hline \boxed{13} \\ \hline\hline \end{array}$$

Project MIND
First - Green

$$\begin{array}{r} 7 \\ +\ 4 \\ \hline 11 \\ \hline\hline \end{array}$$

Project MIND
First - Green

$$\begin{array}{r} 8 \\ +\ 4 \\ \hline \end{array}$$

$$\begin{array}{r} 7 \\ +\ 5 \\ \hline \end{array}$$

$$\begin{array}{r} 7 \\ +\ 5 \\ \hline \boxed{12} \end{array}$$

Project MIND
First - Green

$$\begin{array}{r} 8 \\ +\ 4 \\ \hline 12 \end{array}$$

Project MIND
First - Green

$$\begin{array}{r} 9 \\ +\ 3 \\ \hline \end{array}$$

$$\begin{array}{r} 9 \\ +\ 0 \\ \hline \end{array}$$

$$\begin{array}{r} 9 \\ +\ 0 \\ \hline \boxed{9} \end{array}$$

Project MIND
First - Green

$$\begin{array}{r} 9 \\ +\ 3 \\ \hline 12 \end{array}$$

Project MIND
First - Green

$$\begin{array}{r} 10 \\ -\ \ 6 \\ \hline \end{array}$$

$$\begin{array}{r} 10 \\ -\ \ 5 \\ \hline \end{array}$$

$$\begin{array}{r} 10 \\ -\;\;5 \\ \hline 5 \end{array}$$

Project MIND
First - Green

$$\begin{array}{r} 10 \\ -\;\;6 \\ \hline 4 \end{array}$$

Project MIND
First - Green

$$\begin{array}{r} 10 \\ -\ \ 9 \\ \hline \end{array}$$

$$\begin{array}{r} 10 \\ -\ \ 8 \\ \hline \end{array}$$

$$\begin{array}{r} 10 \\ -\ \ 8 \\ \hline 2 \end{array}$$

Project MIND
First - Green

$$\begin{array}{r} 10 \\ -\ \ 9 \\ \hline 1 \end{array}$$

Project MIND
First - Green

$$\begin{array}{r} 11 \\ - \quad 3 \\ \hline \end{array}$$

$$\begin{array}{r} 11 \\ - \quad 2 \\ \hline \end{array}$$

$$\begin{array}{r} 11 \\ -\ \ 2 \\ \hline 9 \\ \hline \hline \end{array}$$

Project MIND
First - Green

$$\begin{array}{r} 11 \\ -\ \ 3 \\ \hline 8 \\ \hline \hline \end{array}$$

Project MIND
First - Green

$$\begin{array}{r} 11 \\ -\ \ 5 \\ \hline \end{array}$$

$$\begin{array}{r} 11 \\ -\ \ 4 \\ \hline \end{array}$$

$$\begin{array}{r} 11 \\ -\ \ 4 \\ \hline 7 \end{array}$$

Project MIND
First - Green

$$\begin{array}{r} 11 \\ -\ \ 5 \\ \hline 6 \end{array}$$

Project MIND
First - Green

$$\begin{array}{r} 11 \\ -\ \ 7 \\ \hline \end{array}$$

$$\begin{array}{r} 11 \\ -\ \ 6 \\ \hline \end{array}$$

$$\begin{array}{r} 11 \\ -\ \ 6 \\ \hline 5 \end{array}$$

Project MIND
First - Green

$$\begin{array}{r} 11 \\ -\ \ 7 \\ \hline 4 \end{array}$$

Project MIND
First - Green

$$\begin{array}{r} 11 \\ -\ \ 9 \\ \hline \end{array}$$

$$\begin{array}{r} 11 \\ -\ \ 8 \\ \hline \end{array}$$

$$\begin{array}{r} 11 \\ -\ \ 8 \\ \hline 3 \end{array}$$

Project MIND
First - Green

$$\begin{array}{r} 11 \\ -\ \ 9 \\ \hline 2 \end{array}$$

Project MIND
First - Green

$$\begin{array}{r} 12 \\ -\ \ 3 \\ \hline \end{array}$$

$$\begin{array}{r} 11 \\ -\ 10 \\ \hline \end{array}$$

$$\begin{array}{r} 11 \\ -\ 10 \\ \hline 1 \end{array}$$

Project MIND
First - Green

$$\begin{array}{r} 12 \\ -\ 3 \\ \hline 9 \end{array}$$

Project MIND
First - Green

$$\begin{array}{r} 12 \\ -\ 5 \\ \hline \end{array}$$

$$\begin{array}{r} 12 \\ -\ 4 \\ \hline \end{array}$$

$$\begin{array}{r} 12 \\ -\ \ 4 \\ \hline 8 \end{array}$$

Project MIND
First - Green

$$\begin{array}{r} 12 \\ -\ \ 5 \\ \hline 7 \end{array}$$

Project MIND
First - Green

$$\begin{array}{r} 12 \\ -\quad 7 \\ \hline \end{array}$$

$$\begin{array}{r} 12 \\ -\quad 6 \\ \hline \end{array}$$

$$\begin{array}{r} 12 \\ -\ 6 \\ \hline 6 \end{array}$$

Project MIND
First - Green

$$\begin{array}{r} 12 \\ -\ 7 \\ \hline 5 \end{array}$$

Project MIND
First - Green

$$\begin{array}{r} 12 \\ -\ \ 9 \\ \hline \end{array}$$

$$\begin{array}{r} 12 \\ -\ \ 8 \\ \hline \end{array}$$

$$\begin{array}{r} 12 \\ -\ \ 8 \\ \hline 4 \end{array}$$

Project MIND
First - Green

$$\begin{array}{r} 12 \\ -\ \ 9 \\ \hline 3 \end{array}$$

Project MIND
First - Green

$$\begin{array}{r} 17 \\ +\ 40 \\ \hline \end{array}$$

$$\begin{array}{r} 16 \\ +\ 50 \\ \hline \end{array}$$

$$\begin{array}{r} 16 \\ +\ 50 \\ \hline \boxed{66} \end{array}$$

Project MIND
First - Red

$$\begin{array}{r} 17 \\ +\ 40 \\ \hline 57 \end{array}$$

Project MIND
First - Red

$$\begin{array}{r} 29 \\ +\ 10 \\ \hline \end{array}$$

$$\begin{array}{r} 19 \\ +\ 20 \\ \hline \end{array}$$

$$\begin{array}{r} 19 \\ +\ 20 \\ \hline \boxed{39} \end{array}$$

Project MIND
First - Red

$$\begin{array}{r} 29 \\ +\ 10 \\ \hline 39 \end{array}$$

Project MIND
First - Red

$$\begin{array}{r} 32 \\ +\ 12 \\ \hline \end{array}$$

$$\begin{array}{r} 32 \\ +\ \ 6 \\ \hline \end{array}$$

$$\begin{array}{r} 32 \\ +\ 6 \\ \hline \boxed{38} \end{array}$$

Project MIND
First - Red

$$\begin{array}{r} 32 \\ +\ 12 \\ \hline 44 \end{array}$$

Project MIND
First - Red

$$\begin{array}{r} 42 \\ +\quad 7 \\ \hline \end{array}$$

$$\begin{array}{r} 34 \\ +\ 22 \\ \hline \end{array}$$

$$\begin{array}{r} 34 \\ +\ 22 \\ \hline \boxed{56} \end{array}$$

Project MIND
First - Red

$$\begin{array}{r} 42 \\ +\ 7 \\ \hline 49 \end{array}$$

Project MIND
First - Red

$$\begin{array}{r} 42 \\ +\ 36 \\ \hline \end{array}$$

$$\begin{array}{r} 41 \\ +\ 27 \\ \hline \end{array}$$

$$\begin{array}{r} 41 \\ +\ 27 \\ \hline 68 \end{array}$$

Project MIND
First - Red

$$\begin{array}{r} 42 \\ +\ 36 \\ \hline 78 \end{array}$$

Project MIND
First - Red

$$\begin{array}{r} 53 \\ +\ 44 \\ \hline \end{array}$$

$$\begin{array}{r} 53 \\ +\ 42 \\ \hline \end{array}$$

$$\begin{array}{r} 53 \\ +\ 42 \\ \hline 95 \end{array}$$

Project MIND
First - Red

$$\begin{array}{r} 53 \\ +\ 44 \\ \hline 97 \end{array}$$

Project MIND
First - Red

$$\begin{array}{r} 71 \\ +\ 12 \\ \hline \end{array}$$

$$\begin{array}{r} 71 \\ +\ 8 \\ \hline \end{array}$$

$$\begin{array}{r} 71 \\ +\ 8 \\ \hline \boxed{79} \end{array}$$

Project MIND
First - Red

$$\begin{array}{r} 71 \\ +\ 12 \\ \hline 83 \end{array}$$

Project MIND
First - Red

$$\begin{array}{r} 81 \\ +\ 13 \\ \hline \end{array}$$

$$\begin{array}{r} 72 \\ +\ 7 \\ \hline \end{array}$$

$$\begin{array}{r} 72 \\ +\quad 7 \\ \hline \boxed{79} \\ \hline\hline \end{array}$$

Project MIND
First - Red

$$\begin{array}{r} 81 \\ +\quad 13 \\ \hline 94 \\ \hline\hline \end{array}$$

Project MIND
First - Red

$$\begin{array}{r} 47 \\ -\ 24 \\ \hline \end{array}$$

$$\begin{array}{r} 36 \\ -\ 21 \\ \hline \end{array}$$

$$\begin{array}{r} 36 \\ -\ 21 \\ \hline \boxed{15} \end{array}$$

Project MIND
First - Red

$$\begin{array}{r} 47 \\ -\ 24 \\ \hline 23 \end{array}$$

Project MIND
First - Red

$$\begin{array}{r} 58 \\ -\ 17 \\ \hline \end{array}$$

$$\begin{array}{r} 48 \\ -\ 25 \\ \hline \end{array}$$

$$\begin{array}{r} 48 \\ -\ 25 \\ \hline \boxed{23} \end{array}$$

Project MIND
First - Red

$$\begin{array}{r} 58 \\ -\ 17 \\ \hline 41 \end{array}$$

Project MIND
First - Red

$$\begin{array}{r} 73 \\ -\ 22 \\ \hline \end{array}$$

$$\begin{array}{r} 59 \\ -\ 18 \\ \hline \end{array}$$

$$\begin{array}{r} 59 \\ -\ 18 \\ \hline 41 \end{array}$$

Project MIND
First - Red

$$\begin{array}{r} 73 \\ -\ 22 \\ \hline 51 \end{array}$$

Project MIND
First - Red

$$\begin{array}{r} 76 \\ -\ 63 \\ \hline \end{array}$$

$$\begin{array}{r} 74 \\ -\ 21 \\ \hline \end{array}$$

$$\begin{array}{r} 74 \\ -\ 21 \\ \hline \boxed{53} \end{array}$$

Project MIND
First - Red

$$\begin{array}{r} 76 \\ -\ 63 \\ \hline 13 \end{array}$$

Project MIND
First - Red

$$\begin{array}{r} 78 \\ -\ 62 \\ \hline \end{array}$$

$$\begin{array}{r} 75 \\ -\ 20 \\ \hline \end{array}$$

$$\begin{array}{r} 75 \\ -\ 20 \\ \hline \boxed{55} \end{array}$$

Project MIND
First - Red

$$\begin{array}{r} 78 \\ -\ 62 \\ \hline 16 \end{array}$$

Project MIND
First - Red

$$\begin{array}{r} 84 \\ -\ 72 \\ \hline \end{array}$$

$$\begin{array}{r} 80 \\ -\ 30 \\ \hline \end{array}$$

$$\begin{array}{r} 80 \\ -\ 30 \\ \hline \boxed{50} \end{array}$$

Project MIND
First - Red

$$\begin{array}{r} 84 \\ -\ 72 \\ \hline 12 \end{array}$$

Project MIND
First - Red

$$\begin{array}{r} 90 \\ -\ 40 \\ \hline \end{array}$$

$$\begin{array}{r} 85 \\ -\ 30 \\ \hline \end{array}$$

$$\begin{array}{r} 85 \\ -\ 30 \\ \hline \boxed{55} \end{array}$$

Project MIND
First - Red

$$\begin{array}{r} 90 \\ -\ 40 \\ \hline 50 \end{array}$$

Project MIND
First - Red

$$\begin{array}{r} 94 \\ -\ 82 \\ \hline \end{array}$$

$$\begin{array}{r} 95 \\ -\ 72 \\ \hline \end{array}$$

$$\begin{array}{r} 95 \\ -\ 72 \\ \hline \boxed{23} \end{array}$$

Project MIND
First - Red

$$\begin{array}{r} 94 \\ -\ 82 \\ \hline 12 \end{array}$$

Project MIND
First - Red

$$\begin{array}{r} 2 \\ 3 \\ +\quad 7 \\ \hline \end{array}$$

$$\begin{array}{r} 2 \\ 5 \\ +\quad 2 \\ \hline \end{array}$$

$$\begin{array}{r} 2 \\ 5 \\ +\ 2 \\ \hline 9 \end{array}$$

$$\begin{array}{r} 2 \\ 3 \\ +\ 7 \\ \hline 12 \end{array}$$

Project MIND
First - Red

$$\begin{array}{r} 3 \\ 3 \\ +\quad 6 \\ \hline \end{array}$$

$$\begin{array}{r} 2 \\ 6 \\ +\quad 3 \\ \hline \end{array}$$

$$
\begin{array}{r}
2 \\
6 \\
+\ 3 \\
\hline
11
\end{array}
$$

$$
\begin{array}{r}
3 \\
3 \\
+\ 6 \\
\hline
12
\end{array}
$$

Project MIND
First - Red

$$\begin{array}{r} 4 \\ 3 \\ +\quad 4 \\ \hline \end{array}$$

$$\begin{array}{r} 2 \\ 5 \\ +\quad 3 \\ \hline \end{array}$$

$$\begin{array}{r} 2 \\ 5 \\ +\ 3 \\ \hline 10 \\ \hline \end{array}$$

$$\begin{array}{r} 4 \\ 3 \\ +\ 4 \\ \hline 11 \\ \hline \end{array}$$

$$\begin{array}{r} 5 \\ 4 \\ +\ 2 \\ \hline \end{array}$$

$$\begin{array}{r} 4 \\ 1 \\ +\ 3 \\ \hline \end{array}$$

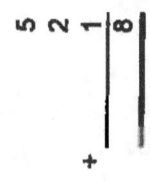

$$\begin{array}{r} 5 \\ 2 \\ +\ 1 \\ \hline 8 \end{array}$$

$$\begin{array}{r} 5 \\ 4 \\ +\ 3 \\ \hline 12 \end{array}$$

$$\begin{array}{r} 6 \\ 4 \\ +\quad 1 \\ \hline \end{array}$$

$$\begin{array}{r} 6 \\ 3 \\ +\quad 2 \\ \hline \end{array}$$

$$\begin{array}{r} 6 \\ 3 \\ +\ 2 \\ \hline 11 \end{array}$$

$$\begin{array}{r} 6 \\ 4 \\ +\ 1 \\ \hline 11 \end{array}$$

www.ingramcontent.com/pod-product-compliance
Lightning Source LLC
Chambersburg PA
CBHW080259180526
45167CB00006B/2594
* 9 7 8 1 5 0 3 2 6 9 2 7 9 *